Biology of Cancer

Biology of Cancer

Second Edition

Michael A. Palladino, Series Editor
Monmouth University

Dorothy Lobo
Monmouth University

PEARSON

Boston Columbus Indianapolis New York San Francisco Upper Saddle River
Amsterdam Cape Town Dubai London Madrid Milan Munich Paris Montréal Toronto
Delhi Mexico City São Paulo Sydney Hong Kong Seoul Singapore Taipei Tokyo

Acquisitions Editor: Michael Gillespie
Associate Editor: Logan Triglia
Marketing Manager: Lauren Harp
Managing Editor: Michael Early
Production Project Manager/
Manufacturing Buyer: Dorothy Cox
Designer: Detta Penna

Full-Service Project Manager: Sreejith Viswanathan, Element LLC
Compositor: Element LLC
Cover Design: Seventeenth Street Design
Cover Image: Steve Gschmeissner/Photo Researchers, Inc.
Printer/Binder/Cover Printer: Courier/Westford

Credits and acknowledgments borrowed from other sources and reproduced, with permission, in this textbook appear on [p. 2, 3, 9, 10, 12, 13, 21, 30,].

Copyright © 2012 by Pearson Education, Inc. All rights reserved. Manufactured in the United States of America. This publication is protected by Copyright and permission should be obtained from the publisher prior to any prohibited reproduction, storage in a retrieval system, or transmission in any form or by any means, electronic, mechanical, photocopying, recording, or likewise. To obtain permission(s) to use material from this work, please submit a written request to Pearson Education, Inc., Permissions Department, 1900 E. Lake Ave., Glenview, IL 60025. For information regarding permissions, call 847/486/2635.

Many of the designations used by manufacturers and sellers to distinguish their products are claimed as trademarks. Where those designations appear in this book, and the publisher was aware of a trademark claim, the designations have been printed in initial caps or all caps.

Library of Congress Cataloging-in-Publication Data

Palladino, Michael Angelo.
    Biology of cancer. — 2nd ed./by Michael Palladino, Dorothy Lobo.
        p. ; cm.
    Rev. ed. of: Biology of cancer/Randall W. Phillis, Steve Goodwin. c2003.
    Includes bibliographical references.
    ISBN-13: 978-0-321-77492-7
    ISBN-10: 0-321-77492-2
    I. Lobo, Dorothy. II. Phillis, Randall W. Biology of cancer. III. Title.
    [DNLM: 1.   Neoplasms.   QZ 200]
    LC   Classification not assigned
    616.99'4—dc23

2011045714

1 2 3 4 5 6 7 8 9 10—**CRW**—15 14 13 12 11

www.pearsonhighered.com

ISBN 13: 978-0-321-77492-7
ISBN 10:   0-321-77492-2

CONTENTS

INTRODUCTION	1
UNDERSTANDING CELL DIVISION	4
The Cell Cycle	4
Unique Properties of Cancer Cells	7
Progression to Tumor Formation	16
WHAT CAUSES CANCER?	18
Role of DNA Damage and Mutations	18
Hereditary Versus Sporadic Cancer	22
SIDE EFFECTS AND DIRTY TRICKS	24
Drug Resistance	24
Cachexia	25
Aneuploidy	25
Differentiation and Cancer Stem Cells	26
Failure of the Immune System and Cancer	27
Epigenetics and Cancer	29
CANCER DIAGNOSIS	30
Common Diagnostic Tests	31
MOST COMMON CANCER TREATMENTS	34
Surgery	34
Radiation	34
Chemotherapy	35
OTHER CANCER TREATMENTS	35
Bone Marrow Transplantation	35

Targeted Therapies	36
Tumor Vaccines	37
Gene Therapy	38
Clinical Trials	38
CONCLUSION	39
WEB RESOURCES	39
BOOKS AND ARTICLES	40

INTRODUCTION

At age 52, Rita decided to retire early from her human resources job. She felt like she was starting to slow down. She seemed more tired than usual. But again, she was transitioning into retirement. She had started to notice some indigestion, and she was feeling a little full after even small meals, but these symptoms were so mild that she barely thought about them and didn't even consider mentioning them to a doctor. A few months after her retirement, Rita made an appointment for her yearly medical exam with her gynecologist. She was stunned when her doctor revealed that her exam had yielded a suspicious mass and sent her to the hospital for an immediate ultrasound. The ultrasound detected a large tumor on her ovary, and Rita was scheduled for a computed tomography (CT) scan followed by surgery 2 days later. Following the surgery, Rita had 6 months of chemotherapy and radiation. She lost her hair and lost some weight, but overall she felt fine and experienced very few side effects from her treatment. After her treatment, her tumor was not detectable on a CT scan, and she was told to return every 6 months to watch for a possible reoccurrence. Eight months later, Rita started to notice a nagging pain in her side that didn't seem to go away. When she returned to her physician, her worst fears were realized as she was told that her cancer had returned and had spread to the other organs in her abdomen. She was referred to a clinical trial at a cancer center 2 hours away. For 3 months, her husband drove her twice a week to the cancer center to get an infusion of an experimental antibody. For the first month and a half, the new treatment appeared to offer hope—her tumors did not shrink, but they did not grow. Her lifestyle did not slow down much—she continued to volunteer at her local animal shelter and babysit her grandchildren every week. Eventually, her tumors progressed despite the treatment, and she ended her participation in the trial. Her health rapidly declined in the last month of her life, and she died in the hospice unit of her local hospital.

Elizabeth Edwards, Michael Douglas, Patrick Swayze, Walt Disney, Mickey Mantle—all have battled cancer (Figure 1). Everyone has been touched by cancer in some way, witnessing a friend, family member, or coworker diagnosed with the disease. The word *cancer* is one that no one wants to hear, and it is usually accompanied by thoughts of pain, hair loss, and poor outcomes. Though in the past a cancer diagnosis was almost always accompanied by hopelessness, many cancers now are very treatable—and in some cases, curable.

The side effects of cancer treatments have been greatly reduced. We are now at a time when earlier diagnosis and new and better treatments are revolutionizing what it means to have a cancer diagnosis. For some forms of cancer, a patient diagnosed today may live several years longer than he or she would have if

FIGURE 1: Walt Disney, Patrick Swayze, and Elizabeth Edwards Are Examples of Celebrities Who Have Battled Cancer. Walt Disney, a long-time smoker, died of lung cancer shortly after his diagnosis in 1966. Patrick Swayze died of pancreatic cancer in 2008, following approximately a year and a half of chemotherapy and participation in a clinical trial for a new drug treatment. Elizabeth Edwards first battled breast cancer in 2004, and it reoccurred while her husband, John Edwards, was campaigning for the presidency in 2007. She succumbed to cancer in 2010.

Source: (left) Moviestore Collection Ltd/Alamy; (center) PhotoNonStop/Glow Images; (right) Everett Collection Inc./Alamy

diagnosed ten years ago. Despite these advances, based on current statistics (Cancer.Gov), it is estimated that 1 in 2 individuals will develop cancer sometime during his or her lifetime, and cancer remains the second-leading cause of death in the United States. Cancer also remains a leading, and increasing, cause of death worldwide. Some cancers, such as the ovarian cancer described in the case above, remain difficult to detect until they are more advanced, and thus, they remain difficult to treat. It is estimated that there will be 1,596,670 new cases of cancer (not including nonmelanoma skin cancers) in the United States in 2011 (Cancer.Gov; Table 1).

The term *cancer* refers to not just one disease, but a class of over 100 diseases characterized, in part, by uncontrolled growth of cells. There is evidence throughout history of the devastation of cancer, from early fossils to descriptions written on papyrus. Hippocrates (460–370 B.C.) is credited with coining the term *cancer*, most likely because the projections often seen extending from a tumor resemble a crab (*cancer* in Latin) (American Cancer Society). In 1971, President Richard Nixon declared a "War on Cancer," signing into law the National Cancer Act, which greatly increased the amount budgeted to study cancer. Though progress has been made in the diagnosis and treatment of many cancers, some types of cancer remain extremely deadly. To some degree, there is now a renewed interest in cancer research, spurred in part by celebrities such as cyclist Lance Armstrong and actress Farrah Fawcett, who have spoken out publicly about their battles with cancer. In 2009, Senator Edward Kennedy (who would later succumb to brain cancer) introduced a bill called the 21st Century Cancer ALERT (Access to Life-saving Early detection, Research, and Treatment) Act, which was intended to revitalize cancer research and increase patient access to prevention and treatment options. This bill did not make it through the committee process before that session of Congress ended, and would need to be reintroduced. At a time when health-care costs are being debated, it is critical to understand both the impact of cancer on society and the causes and potential treatments of these diseases.

This booklet provides you with an introduction to cancer as a disease, including causes of cancer and cancer treatments. Throughout, key terms appear in bold to help you learn important concepts related to the biology of cancer. At the back of the booklet is an excellent collection of references and web links that you should consider if you are seeking current information on a cancer topic of interest.

TABLE 1: Leading Sites of Cancer.

Incidence per 100,000, 2004–2008	Male	Female
All sites	541.0	411.6
Lung & bronchus	75.2	52.3
Prostate (male), Breast (female)	156.0	124.0
Colon & rectum	55.0	41.0
Pancreas	13.6	10.7
Leukemia	16.1	9.7
Non-Hodgkin lymphoma	24.0	16.5
Esophagus	7.8	1.9
Liver & intrahepatic bile duct	11.2	3.9
Ovary	—	12.8
Urinary bladder	37.5	9.2
Uterine corpus	—	23.4
Kidney & renal pelvis	20.0	10.2
Brain & other nervous system	7.7	5.4

Source: http://seer.cancer.gov/csr/1975_2008/browse_csr.php?section=1&page=sect_01_table.04.html

UNDERSTANDING CELL DIVISION

Cancer cells divide out of control. In Rita's case, one cancerous cell rapidly divided to eventually form a large tumor in her ovary. Then cells from the initial tumor "escaped" and spread to other organs, continuing to divide to form secondary tumors. To understand how cell growth becomes out of control in cancer, it is important to understand how the growth of cells is regulated under normal conditions. Most body cells that make up our tissues and organs do not divide continuously, but rather only when needed. For example, when you get a paper cut on your finger, skin cells that are alive, but not actively in the process of cell division, will begin to divide again to close the wound. The process of cell division—when and how often a cell will divide—is tightly regulated. Many conditions, both external and internal, must be met before a cell can divide. Internally, a cell must grow in size and content before it divides. Enough proteins, membrane, organelles (cellular compartments, such as mitochondria, ribosomes, and the nucleus), and carbohydrates must be produced so that, when those materials are divided in half in cell division, there is enough to support each of the two daughter cells formed. Externally, cells must be in contact with neighboring cells and be properly anchored in place to appropriately form the needed tissue.

To divide, cells must also receive a complex array of signals transmitted by nutrients, hormones, and growth factors released by other cells. Some signals received are also negative—inhibiting or blocking cell division. Most cells in the body are not dividing much at all, and many cell types have undergone terminal differentiation and matured (becoming a specific cell type with a specific function), never to divide again. The signals these mature cells receive from neighboring cells, hormones, and other cues prevent them from returning to a state of cell division. Cancer cells escape these normal controls of cell cycle regulation.

The Cell Cycle

All of the internal and external signals that regulate cell division are integrated to regulate the **cell cycle**, the sequence of events that are repeated each time a cell grows and divides to form two daughter cells (Figure 2).

As the cycle proceeds, the cell goes through interphase, followed by mitosis. During interphase, the cell passes through a growth phase (G1), followed by a DNA synthesis phase (S), during which all of the cell's DNA is copied, and a second growth phase (G2). At the end of G2, the cell enters into mitosis (M phase), where division of the cell into daughter cells occurs. *Checkpoints* at the transition from each phase of the cell cycle to the next are tightly regulated by a

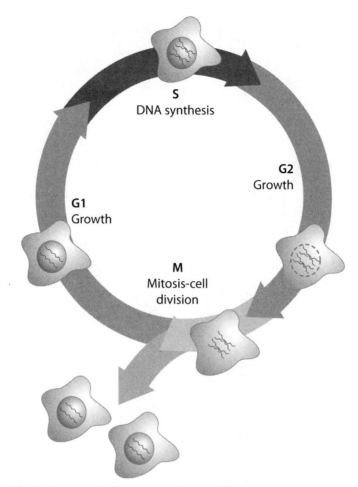

FIGURE 2: The Cell Cycle. Cells go through four different stages each time they divide. Cells start with a growth phase, called G1, in which they synthesize proteins, lipids (fats), and carbohydrates and perform important metabolic functions. Once they grow properly and receive signals that stimulate cell division, they move to the S phase, in which they replicate their DNA. After DNA replication, further growth and metabolism occur in the G2 stage. The cells also check the chromosomes for errors and make any necessary repairs to damaged DNA. Finally, the cells undergo mitosis, the M phase, in which chromosomes are divided evenly and split into two genetically identical cells. Each of those cells can then cycle, or go through the process, again.

system that combines signals from both internal and external cues into a single control switch for cell division. At the start of the cell cycle, the concentration of proteins called *cyclins* is very low (Figure 3). As signals are received to stimulate cell division, cyclins accumulate to a high concentration within the cell. When cyclins become more concentrated, they combine with a second class of proteins

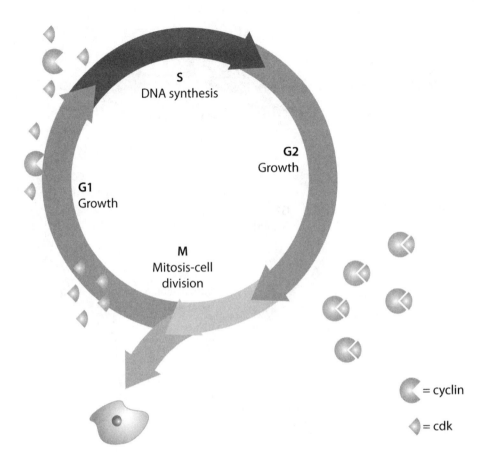

FIGURE 3: Control of Cell Cycle Checkpoints. Early in the cell cycle, cyclins are at a very low concentration in the cell. When growth factors signal the cell, it stimulates cyclin production. When the cyclins are present in a high concentration, they bind with the cyclin-dependent kinases, or cdk's. These two-molecule combinations, or dimers, then signal other molecules that allow the cell cycle to proceed. Once the cell cycle proceeds, the cyclins that were at a high concentration are uncoupled from the cdk's and destroyed, and the cdk's are deactivated. There are different cyclin-cdk complexes used at each stage of the cell cycle, and they are activated and deactivated in the appropriate stage. For simplicity, cyclin-cdk's involved in the progression through the M phase are diagrammed here.

called *cyclin-dependent kinases*. When cyclin and its corresponding cyclin-dependent kinase bind to each other, they become active enzymes that stimulate additional proteins. These target proteins function to actually move the cell from one phase of the cell cycle to the next. For example, many of the targets of cyclin-dependent kinases activated at the G1 checkpoint are enzymes required for DNA synthesis. Thus, the DNA replication machinery used in the S phase is engaged actively only after all of the signals from G1 phase cues have combined to produce high G1 cyclin levels. These integrated signals are combined into a unified activation event triggered by the G1 cyclin-dependent kinase enzyme complexes.

Each checkpoint at each transition in the cell cycle is regulated and triggered by specific sets of cyclins and their corresponding kinases. The production of cyclins at each stage is stimulated by positive growth signals and inhibited by negative growth signals. Through this give-and-take, the cell cycle can be either stimulated, allowing the cell to divide, or stalled, arresting the cell in its existing growth phase. The accumulation of positive signals, in essence, is akin to stepping on the gas pedal of a car to move it forward. The accumulation of negative signals is like stepping on the brake pedal to stop the movement of a car. To drive appropriately, the correct balance of using the gas pedal and using the brake pedal is needed. If the gas pedal is depressed too long or the brake pedal is not used, accidents can happen. For the cell, having the cell cycle progress to cause cell division when it is not needed is also a real problem. Cancer cells often have defects in multiple cell cycle regulation genes, contributing to excessive cell division.

Unique Properties of Cancer Cells

Unlike infectious diseases caused by bacteria or viruses, cancer is not "caught" from an outside source, and it is not "contagious." Cancer arises in an individual's body from one faulty cell that undergoes many genetic changes. The hundreds of types of cancer that have been identified are all unique; many different types of genetic changes can result in cancer and thus influence treatment strategies, as will be discussed in a later section. However, all cancer cells have many characteristics in common—most obviously, uncontrolled cell division. Cancer is "cell division gone bad." Over a decade ago, scientists Douglas Hanahan and Robert Weinberg (2000) described the uncontrolled cell division of cancer cells as being fueled by six distinct features, the "hallmarks" of cancer that separate them from normal cells:

1. Growth without "go" (positive) signals,
2. Failure to respond to "stop" (negative) signals,
3. Evasion of programmed cell death (apoptosis),
4. Unlimited cell division,
5. Sustained angiogenesis (stimulation of blood vessel growth), and
6. Tissue invasion and metastasis.

These hallmarks of cancer are caused by mutations, damage to DNA, that lead to the incorrect function of proteins that are involved in these processes. For each of these six hallmarks of cancer, there are a multitude of genes that may be involved. Some of the concepts of these hallmarks and key examples will be highlighted in the sections that follow.

Growth Without "Go" Signals

In order to divide, cells must receive "permissive" signals from their environment in the form of **growth factors**. Growth factors are often proteins that are naturally found in the body and attach to receptors on the surface of cells, permitting them to carry out cell division. Cancer cells can divide without having growth factors present. This ability results from genetic mutations found in proteins that control "go" signals. Such proteins are encoded by genes called **oncogenes** and are compared to the gas pedal of a car. If operating correctly, they help you "go" forward appropriately. When mutated, these proteins cause the cell cycle to progress in an uncontrolled fashion (like having a gas pedal stuck in the acceleration position). For example, some oncogenes may be cell surface receptors for growth factors that are mutated in such a way that they assume the shape that they would have if a growth factor was bound to them even though such an interaction has not occurred. This damage causes the receptor to send "divide" signals to the rest of the cell incorrectly. Alternatively, some cancers have mutations in the proteins inside the cell that are activated by the receptor and transduce or transmit the signal into the cell to activate cell division. An example is a gene called *ras*, which encodes a protein that functions as a signal transduction molecule. The mutant ras protein becomes overactive in cancer cells and delivers a growth signal even when none is actually being transmitted from outside the cell.

Among the most devious of the systems that lead to the mistaken signaling of growth factors are cancers that co-opt neighboring healthy cells in order to overexpress these growth factors. This effectively represents a hijacking of healthy cells by cancer cells to support their rapid growth. Cancer cells that thrive on extra growth signals get their "fix" by releasing molecules that affect nearby cells and fool them into expressing high levels of growth factors. With this in mind, the complexity of cell types within tumors can make sense. Many tumors are a mixture of cancer cells that are capable of rapid growth, invasive growth, and metastasis and noncancer cells that may function to "support" the rapid growth of the cancer cells.

Failure to Respond to "Stop" Signals

Cell division is tightly controlled; if division is not needed, "stop" signals in the cell's environment will prevent division. For example, lack of space in which to divide is one such environmental "stop" signal. When normal cells are in contact with each other and there is no space between them, they do not divide. This is called **contact inhibition**. Cancer cells do not follow the rules of contact inhibition—in the absence of space, they will continue to divide, overcrowding their environment and forming tumors (Figure 4). As tumors become larger, they often press on other organs or nerves due to the lack of space, and this can cause pain. "Stop" signals are lost due to mutations in proteins that prevent the cell cycle from progressing. These proteins, called **tumor suppressors**, can be compared to the brakes of a car. When working appropriately, they stop forward

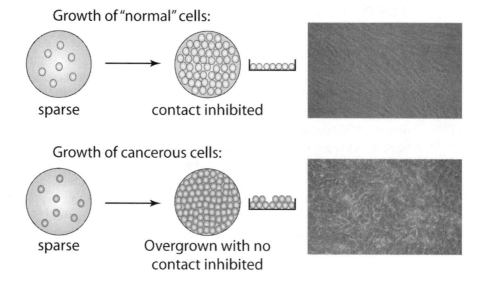

FIGURE 4: Cancer Cells Lack Contact Inhibition. This phenomenon can be easily seen when cells are grown in culture dishes in the laboratory. Normal cells, when plated sparsely, will continue to proliferate until the entire dish is filled and then, once they are touching each other, will stop dividing. As shown in the top photo, normal fibroblasts, a type of connective tissue cell found in skin, line together and form a single layer on the bottom of a culture dish. In contrast, cancer cells, when plated sparsely, will continue to proliferate and will overgrow in the dish, piling up on top of each other—contact inhibition is lost in cancer cells. The bottom photo shows cancerous fibrosarcoma cells—their organization in the dish is more random, and in some areas, cells have begun to pile on top of each other.
Source: Photos by D. Lobo (Monmouth University).

movement of a car. When the brakes fail, the car will keep moving. Similarly, if tumor suppressors are mutated, the cell cycle will progress when it should not.

Evasion of Programmed Cell Death

When a cell's DNA is damaged, the cell cycle checkpoints will be activated to "stop" the cell cycle, and mechanisms that the cell contains to repair the damage will be employed. Most often, the cell's repair mechanisms will be able to fix the damage to the DNA, and the cell cycle will be resumed. Sometimes, however, the DNA damage that is accumulated is so overwhelming that it becomes impossible for the cell to fix it. In these cases, a mechanism called **programmed cell death** or **apoptosis** is triggered, which causes the cell to systematically kill itself (Figure 5). This cellular suicide is highly controlled—enzymes are triggered that systematically chop up the proteins, DNA, and organelles in the cell and neatly package them in membrane-bound sacs that can be easily engulfed by cells of the immune system. By the time a cell is cancerous, it has accumulated a significant amount of damage and yet has not triggered apoptosis. In addition, the metabolic stresses of rapid cell division, including oxygen deprivation, normally trigger cell death, except in cancer cells. Finally, apoptosis is triggered when cells lose contact with other cells and their anchors are disrupted, except in cancer cells. Cancer cells have mutations in proteins that are needed to trigger the process of apoptosis.

The biochemical pathways that trigger apoptosis are becoming more clearly understood and are quite complex. There are several mutations that can disrupt this process and allow cancer cells to survive despite the presence of several apoptosis-triggering signals. An example of one such protein, which has been found to be mutated in over half of human cancers, is p53. p53 normally

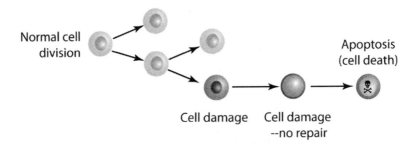

FIGURE 5: Apoptosis. Damaged cells can be repaired, or they can undergo apoptosis if the damage is so severe that it cannot be repaired. Sometimes damaged cells that should undergo apoptosis survive—this can lead to the formation of cancer.
Source: www.cancer.gov/cancertopics/cancerlibrary/what-is-cancer.

functions as a tumor suppressor—when DNA becomes damaged, p53 halts the cell cycle by blocking the activity of cyclins in the G1 phase so that the damage can be repaired. If the damage is too extensive, p53 can trigger apoptosis. So, as a tumor suppressor, p53 stops growth, but when this protein is damaged or missing, cells that should be undergoing repair or apoptosis are incorrectly allowed to continue to divide. This failed repair is especially dangerous if the defects in the cell lead to increases in cell division rates, the increased ability of cells to penetrate tissues by invasive growth, or the ability of cells to break free and spread to distant sites in the body. When these defects are present and not repaired, the result is an aggressive cancer that is rapidly growing and likely to spread. **Therefore, either the loss of cell cycle growth control or the loss of death signals can have a similar outcome: unregulated cell growth.**

The loss of apoptotic capacity in cancer cells has important implications for cancer treatment. Many cancer treatments in wide use today, including radiation and chemotherapy, are designed to cause DNA damage to rapidly dividing cells. The logic of these treatments is that targeting DNA synthesis will focus the treatment on cells undergoing rapid division. Therefore, cancer cells will be selectively affected, and slower-growing healthy cells will experience less damage. The catch is that, even though the rapidly dividing cancer cells accumulate damage from the treatment, they frequently fail to react to that damage by triggering cell death. Therefore, the very design of the treatment is often defeated by one of the properties that are the hallmarks of cancer. The good news in this regard is that research on the mechanisms of apoptosis is revealing that it is quite complex and can be triggered many different ways. Though some cancers may have mutations that block apoptosis from damage to DNA, they may still activate cell death from loss of cellular anchors. A number of cancer treatments are being developed to target the apoptosis pathways that are still intact in cancer cells in order to activate cell death and eliminate cancer cells. When patients are treated with chemotherapy, the drug is typically found dispersed widely throughout the body. When a drug is used in this way to trigger apoptosis in unanchored cells, can you think of some unintended consequences? There are many challenges to providing effective, yet tolerable, cancer treatments.

Unlimited Cell Division

Normal cells can divide approximately 60 times. This is due to the difficulty in replicating the ends of double-stranded DNA, called **telomeres**. During DNA replication, an enzyme called DNA polymerase is needed to copy each strand of DNA. DNA polymerase can copy DNA only by adding new subunits of DNA (called *nucleotides*) to an existing chain of nucleotides. Because of a quirk in the way DNA polymerase adds nucleotides when making a strand of DNA, it is impossible for DNA polymerase to add new nucleotides to these ends, causing the DNA to become shorter and shorter with each round of division. Telomere sequences are repetitive stretches of DNA that do not code for any proteins, so losing some of these sequences throughout the lifetime of a cell is permissible. Eventually, after about 60 divisions, the DNA becomes so short that the cell will not divide anymore.

In comparison, cancer cells can divide infinitely. This difference is due to the activation of a protein called **telomerase** in cancer cells that allows them to permanently carry out DNA replication. The job of telomerase is to add nucleotides to the telomeres, making them longer (Figure 6). Because of this, the

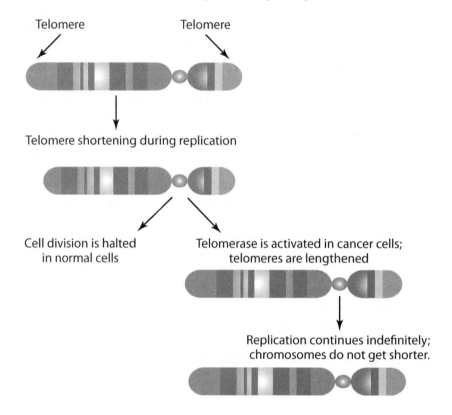

FIGURE 6: DNA Replication Results in the Loss of Some Telomere Sequences. Normal chromosome ends, called telomeres, have repeated sequences. During replication, the telomeres become shorter. When telomeres become too short, cell division is halted. However, cancer cells have telomerase, an enzyme that is able to lengthen telomeres, allowing replication to continue indefinitely.
Source: Adapted from http://10.1146/annurev-genom-082908-150046.

cell will be able to continue to divide despite many rounds of cell division—as the DNA starts to become shorter, telomerase will make it longer again. Telomerase is active early in development so that rapid division of a new embryo can be accomplished. Cells that undergo constant division, such as stem cells and sperm and egg cells that have the potential need for massive amounts of cell division, will have active telomerase. Telomerase is not active in the vast majority of normal cells, and in an adult, very few cells have active telomerase. However, approximately 90% of cancers have telomerase present, due to mutations that cause the production of this protein when it is not appropriate to have it present in the cell.

Sustained Angiogenesis

Angiogenesis is the formation of new blood vessels. Cancer cells are rapidly growing and so require a rich supply of nutrients and oxygen and a means to eliminate waste products. The growth of blood vessels in a tumor is key for its ability to continue growth. Without an adequate blood vessel supply to "feed" the tumor, the tumor would not be able to grow more than a few millimeters in diameter. However, the addition of blood vessels makes it possible for large tumors to be sustained (Figure 7). The production of blood vessels is stimulated by the tumors—cancerous cells release proteins that trigger existing blood

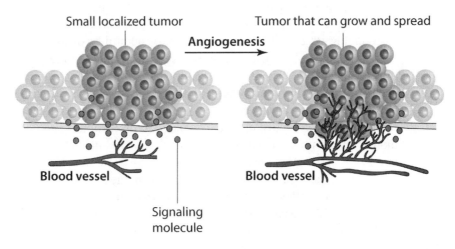

FIGURE 7: Tumors Are Capable of Angiogenesis. Cancer cells release proteins, including VEGF and bFGF, that stimulate blood vessels to grow. Tumors that have adequate blood supplies due to new vessels can continue to grow and spread.
Source: http://www.cancer.gov/cancertopics/understandingcancer/angiogenesis/page3.

vessels to grow. Many angiogenic proteins have been identified, but the two most critical are vascular endothelial growth factor (VEGF) and basic fibroblast growth factor (bFGF). Both VEGF and bFGF are overproduced by the tumor cells and secreted into the environment of the tumor. When these proteins reach receptors on the endothelial cells of blood vessels, they stimulate the endothelial cells to divide to form new vessels. Oftentimes the production VEGF and bFGF can be at such a high level in tumors that hypervascularization occurs. Not only does this supply the cancer cells with the nutrients and oxygen they require for their rapid growth, but also in advanced cancers, these tumors can rob surrounding healthy tissue of adequate nutrients and oxygen and contribute significantly to the pathology of cancer. Highly vascularized tumors tend to bleed profusely when disturbed during biopsy procedures or surgical removal.

As we will discuss later in the booklet, inhibitors of angiogenesis have been studied as potential tools in treating cancer. Since they prevent angiogenesis, they prevent the growth of tumors and thereby keep the tumor "stable"—they do not kill the tumor cells. For this reason, they are used in combination with other treatments. Can you anticipate potential problems with this kind of treatment?

Tissue Invasion and Metastasis

About 90% of the deaths associated with cancer occur due to the spreading of a single tumor to other locations in the body, which is termed **metastasis**. In order for a cancer to spread, the tumor must first have the ability to invade the surrounding tissue. This is usually due to the ability of the cancer cells to lose adhesion molecules that normally keep cells stuck together and to exhibit motility. Usually, the loss of attachment to other cells triggers apoptosis, but cancerous cells have accumulated other mutations, as discussed earlier, that prevent apoptosis from occurring. Cancerous cells are capable of crawling by rearranging their cytoskeleton to send out projections, called *filopodia*. Such motility is common in cells early in development, but is activated again in cancer due to the misregulation of genes that control the cytoskeleton. Once migration of a cancer cell takes place, the tumor can metastasize if the cancerous cell crosses into the bloodstream or lymphatic system, survives transport in the blood, and successfully leaves the bloodstream and reattaches in a new location (Figure 8). The likelilood of a cancer cell surviving this journey is low. However, a large tumor can shed incredibly large numbers of cells into the bloodstream on a daily basis, making it likely that survivor cells may set up residence in another organ.

Eventually, the tumor cells can get "stuck" in the capillaries of distant organs. Each organ contains cells that have different varieties of adhesion molecules expressed on their surfaces—specific types of cancerous cells therefore tend to "stick" preferentially to specific organs. For example, breast cancer cells have

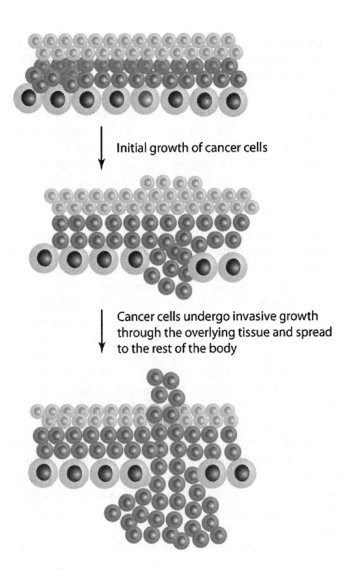

FIGURE 8: Invasive Growth of Cancer Cells. Early in tumor development, cancer cells grow, but remain in place in the tissue. As cancer progresses, the cells develop the ability to change shape, secrete enzymes that break down surrounding tissues, and grow through adjacent cell layers. This invasive growth allows cancer cells to metastasize, or spread, to new locations in the body where they can attach and form new tumors.

adhesion molecules that are also seen in lung tissue. Therefore, once breast cancer cells metastasize, the new secondary tumors are often in the lung. A tumor that has metastasized contains cells that have the same characteristics as the primary tumor. So, even though the tumors are found in the lung, at a cellular level they have the same shape, genetic make-up, and characteristics of the primary breast cancer tumor. The presence of metastatic tumors often interferes with the function of the host organ, further complicating the treatment of cancer. In addition, the presence of metastasis may not be observed immediately. In fact, metastatic cells can survive in distant organs potentially for months or years after the original cancer is eradicated. For this reason, most of the time a "second" cancer is found in a person previously treated for cancer, the origin of the tumor is determined to be a metastatic tumor from the primary cancer.

Progression to Tumor Formation

Cancer is caused not by a single genetic change in a cell, but rather by multiple genetic changes in combination that allow growth to become unregulated and the hallmarks of cancer to be realized. Since we receive one copy of each chromosome from each parent, we have two copies of every gene. In the early 1970s, a two-step model for cancer development proposed that cancer arises when both copies of an oncogene, or tumor suppressor, must be mutated in order for cancer to occur. The first step, *initiation*, is the first genetic mutation in an oncogene, or tumor suppressor, that predisposes a cell to cancer. The second step, *promotion*, is the mutation of the second copy of the same gene, causing the function of that gene to be totally lost. Since this model was proposed, the genetics of specific types of cancer have been better studied, and it is now known that mutations in many genes, possibly 10 or more, must occur in the same cell lineage over time in order to cause *progression* of a normal cell to a cancerous one (Figure 9). Initially, a single cell may accumulate one or two mutations that cause cell division to increase. This may give rise to a patch of tissue that forms a benign tumor—an area that exhibits too much growth, but that is not yet capable of invading other tissues and does not show any of the other hallmark signs of cancer.

Additional mutations must occur in this cell in order to cause the daughter cells to divide faster and faster and to spread to other tissues and cause angiogenesis. For example, if we follow a cell lineage from its healthy start to its final fate as a part of a fully developed cancerous tumor, we might observe the following mutation sequence: Initially, the cell might acquire a mutation in growth factor receptors that would increase the growth-stimulating signals and decrease the growth-inhibiting signals. The descendants of this cell will now divide faster than neighbors and form a small benign tumor. Next, a cell within this tumor might develop an additional mutation that blocks apoptosis so that the stress of increased growth rates will not trigger cell death. Cells derived from this mutant cell not only will divide faster, but also will not undergo apoptosis and disappear from the tumor. Within this cell-death-resistant population of tumor

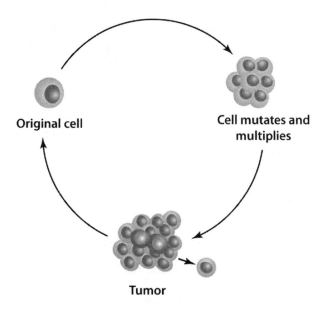

As more mutations occur, cells spread to other locations, causing metastasis.

FIGURE 9: Progression of a Normal Cell to Form a Cancer Tumor. Initial mutations can cause unregulated growth. Additional mutations can lead to tumors and cause progression to metastasis.

cells, a mutation might occur that stimulates the release of VEGF and causes the formation of blood vessels that will supply the tumor with nutrients. Again, further mutations might occur in a cell within the tumor that will activate cell motility and allow that cell and its descendants to lose anchors and move from the original site. When these cells move to new sites in the body, they will retain all of these mutations: They will be capable of growing rapidly, be resistant to cell death, and be able to induce the formation of new blood vessels. Finally, cells in this population might mutate further to activate telomerase and become immortal, unconstrained by the limits on DNA replication imposed by telomere shortening.

Mutations are a rare occurrence, so it takes some time for multiple mutations to change a normal cell to a cancerous one. This is consistent with the observation that cancer is largely a disease of old age. Time is required for mutations to accumulate, although factors that increase the rate at which mutations occur will accelerate the appearance of cancer cells in the body. Many cancer prevention strategies today focus on the early detection of cancer. If pre-cancerous tumors can be detected and removed before additional mutations make them more deadly, cancer can be prevented. This concept is the basis for the use of routine mammograms to detect early breast cancers, colonoscopies to remove pre-cancerous growths in the colon, and Pap smears to detect cell growth changes in the cervix.

WHAT CAUSES CANCER?

Role of DNA Damage and Mutations

It is clear that cancer is caused not by a single event, but by multiple problems occurring in a population of cells that transform them from behaving normally to having unregulated growth. Cell behavior is ultimately controlled by the proteins that are produced using the cell's DNA. **Mutations** in DNA alter the ability of proteins to be made correctly; when proteins involved in cell cycle regulation are missing or incorrect, there can be dramatic consequences for a cell. A mutation is defined as a change in the genetic sequence of DNA. As mentioned previously, DNA is made up of nucleotides. Nucleotides contain one of four different subunits called **nitrogenous bases**: adenine (A), guanine (G), cytosine (C), and thymine (T). It is the order of A's, G's, C's, and T's in a DNA sequence that serves as the blueprint to construct a protein. When this order is altered, proteins are not made correctly. For example, a DNA strand with the sequence ATGCCCAGA means something very different from a strand with the sequence ATGACCAGA.

Mutations can happen spontaneously during DNA replication, or they can be the result of damage to DNA. Spontaneous mutations occur when DNA polymerase adds incorrect nucleotides during DNA replication—for example, when an A-containing nucleotide is added instead of a C-containing nucleotide. DNA polymerase "proofreads" during replication, so mistakes are rare (estimated at one in a billion when studied in bacteria), and even when such mistakes occur, they are usually recognized by DNA repair enzymes in the cell and fixed. Spontaneous mutations that escape the cell's proofreading and repair mechanisms are problematic.

Mutations can also arise due to **DNA damage**. DNA damage is defined as a structural modification to the DNA strand that affects the shape or double-helical arrangement of the DNA. If the DNA's shape is altered, its ability to be replicated correctly is often affected, causing mutations to occur. Once DNA damage occurs, there are many DNA repair mechanisms that can be activated in the cell, and as mentioned before, the cell cycle can be halted at the checkpoints to give time for the damage to be repaired. Usually, when the damage is so extensive that it cannot be repaired, the cell will undergo apoptosis. However, sometimes cells with DNA damage escape apoptosis—when this occurs, damage can lead to mutations during DNA replication.

Substances that lead to cancer are called **carcinogens**. Carcinogens and some examples of substances falling into these categories include:

> Radiation: X-rays, CT scans, radon, and ultraviolet (UV) radiation;
>
> Chemicals: mustard gas, some hair dyes, and chemotherapy drugs themselves; and
>
> Environmental toxins: tar from cigarette smoke, coal, some hormones, dietary fat, and infectious agents.

Many of these agents act directly to damage DNA, although some, such as hormones, work by promoting growth.

Radiation

Radiation promotes cancer by triggering the production of **free radicals**—atoms with unpaired electrons. Such atoms can cause structural damage to both DNA and proteins, including breaks in chromosomes. There is some evidence that **antioxidants**, such as beta-carotene, lycopene, and some vitamins, may help to protect cells against free radicals, but this is still an area of active research. Radiation can be found in a variety of environmental sources. UV radiation from the sun has been linked to the development of skin cancer by directly causing DNA damage. **Radon** is a radioactive gas produced naturally in the environment as uranium breaks down in soil. It can reach high concentrations in homes and buildings, where it enters through the foundations. According to the U.S. Environmental Protection Agency (EPA), radon is the leading cause of lung cancer among nonsmokers. Most new homes are now tested for radon levels, and if they are found to be high, fan systems can be installed to help remove the indoor buildup of radon. X-rays and CT scans, used for medical diagnosis, have also been linked to cancer through the production of DNA damage. Repeated CT scans, which utilize much more radiation than X-rays, may increase the risk of developing cancer later in life, especially when they are performed on children. A study published in the *New England Journal of Medicine* (Brenner & Hall, 2007) estimated that 0.4% of all cancers in the United States may have been caused by CT scans.

Chemicals and Environmental Toxins

Many of these agents cause cancer by directly damaging DNA, leading to mutations. **Mustard gas** was used in chemical warfare in World War I, resulting in lower white blood cell counts in soldiers that were exposed to it and, in subsequent years, an increase in the cancer death rate in these exposed soldiers. When DNA is exposed to mustard gas, damage occurs on the A and G nucleotide subunits, which disrupts DNA replication and can cause mutations along the DNA that lead to cancer if not repaired appropriately. Interestingly, after World War I, mustard gas became one of the first **chemotherapy** agents used to treat cancers of white blood cells, and patients receiving this therapy did show some temporary improvement.

Similarly, coal, tar, asbestos, and other chemical agents that cause DNA damage may trigger mutations that can lead to cancer. Sometimes it seems impossible to avoid the many chemicals that we are exposed to in modern life that may lead to DNA damage. However, one of the most studied causes of cancer, smoking cigarettes, is one that can be avoided. For a very long time, epidemiological evidence has been accumulating that links cigarette smoking to lung cancer. (Epidemiological evidence is simply data that draw a correlation between the occurrence of a disease and other factors.) As the prevalence of smoking in a population increases, the prevalence of lung cancer also increases. In the United States, this was true when men began to smoke in large numbers and much later when women began to smoke in large numbers. There is also a relationship between the number of cigarettes smoked and the likelihood of developing cancer. Finally, stopping smoking can be demonstrated to decrease the likelihood of developing lung cancer. This is all strong evidence, but none of it demonstrates a direct cause-and-effect relationship between cigarette smoking and cancer. Nevertheless, the U.S. Food and Drug Administration (FDA) mandates that warnings must be placed on cigarette packages (Figure 10).

For years, the tobacco companies obscured the dangers of smoking by attacking the fact that only correlations between smoking and lung cancer were used to argue that smoking actually causes cancer. It would be better to show cause and effect directly in order to demonstrate that smoking cigarettes can cause lung cancer. The first evidence for a direct cause-and-effect relationship came in 2001 when researchers were able to demonstrate that a chemical in cigarette smoke causes mutations in the tumor suppressor gene that encodes the p53 protein. The chemical, benzopyrene, is found in the tars of cigarette smoke. Ironically, in an attempt to rid the body of this insoluble chemical, the liver converts benzopyrene into the more chemically reactive benzyopyrene diol epoxide. It is this chemical that interacts with DNA and causes mutations in specific sites in the p53 gene. As we have already learned, the p53 protein plays

FIGURE 10: Cigarette Warnings. Warning labels declaring the risks of smoking must be included on cigarette packages. The Family Smoking Prevention and Tobacco Control Act, passed in 2009, mandates that 50% of the front and back of each package must describe the negative impacts of smoking. New packages must adhere to this rule in 2012.

Source: U.S. Department of Health and Human Services.

an important role in arresting cell division, activating repair enzymes, and, if necessary, triggering cell death. The loss of these tumor-suppressing functions is an important step in the development of lung cancer.

Benzopyrene is not the only carcinogen in cigarette smoke. Cigarette smoke contains approximately 7,000 chemicals, 69 of which are known carcinogens, including arsenic, cadmium, and polonium-210 (a radioactive element, which, interestingly, was used to murder former Russian KGB agent Alexander Litvinenko in Britain in 2006). All of these chemicals may have a cumulative effect in causing substantial DNA damage and mutations. In addition, there tends to be a lag of 20–30 years between the time an individual stops smoking and the time lung cancer appears. We know that cancers develop through the accumulation of a series of mutations. Even if some of these mutations are inherited, it takes time for other mutations to be caused, by smoking and other factors, before cancer can develop.

Viral Infections and Cancer

Infection with a variety of viruses has been linked to an increase in cancer risk. Viruses that have been associated with causing cancer are called **oncoviruses**. It is believed that viruses can help promote cancer because the genes of the infecting virus often combine with the genes of the host (the human cells); this can alter the regulation of the host cell's genes, allowing growth-stimulating genes to be activated. Additionally, some viruses carry their own genes that, when introduced to human cells, can cause the human cells to undergo rapid growth.

One of the first viruses linked to cancers in humans is the **Epstein-Barr virus (EBV)**. EBV causes the disease mononucleosis and has been shown to slightly increase the risk of lymphoma (cancer of the lymphatic system). **Human papillomavirus (HPV)** has been identified as the cause of most types of cervical cancer. The U.S. FDA has recently approved two different vaccines, Gardasil and Cervarix, to protect against the two strains of the virus known to cause 70% of cervical cancer cases. These are just two examples of several viruses that are known to have a link to cancer (Table 2). Interestingly, in mice, the **mouse mammary tumor virus (MMTV)** has been shown to cause the majority of breast tumors in mice, although a definitive link between viruses and human breast cancers has not been established. As our understanding of human viruses continues to grow, we will more than likely uncover more links between infections and cancers.

Hereditary Versus Sporadic Cancer

We have seen that there is a close link between mutations and the development of cancer. In fact, our current understanding of cancer suggests that all cancers arise as a result of mutations in cells. When a cancer cell divides, it passes its mutations on to the two new cancer cells that are formed. However, saying that all cancers arise as a result of mutations in cells is not the same as saying that all cancers are hereditary. Mutations that arise in pancreatic cells may cause pancreatic cancer in that individual. But those mutations are in the cells of the

TABLE 2: Viruses Associated With Specific Cancers.

Virus	Cancer Type
Hepatitis B and C	Liver cancer
Human papillomavirus (HPV)	Cervical cancer
Epstein-Barr virus (EBV)	Lymphatic cancer
Kaposi's sarcoma herpes virus	Kaposi's sarcoma (a type of skin cancer)
Human T-lymphotropic virus Type I (HTLV-1)	T-cell leukemia

pancreas; they are not in the egg or sperm cells that supply genetic information to the next generation. Egg and sperm cells are **germ line cells**. Pancreatic cells are among the cells known as **somatic cells**, or body cells. The genetic information from somatic cells is never passed to a new generation. Thus, mutations in somatic cells may lead to cancer, but these mutations cannot be inherited and cause cancer in the patient's children.

We do know that some forms of cancer can be inherited. A good example is some forms of breast cancer. Breast cancer sometimes will "run in families," and the affected families will have a very high incidence of breast cancer. But even in these high-risk families, not every woman in the family is certain to develop breast cancer. What is inherited is a breast cancer *predisposition*, or an increased likelihood of developing breast cancer. In light of what we know about cancer, what can explain this inherited predisposition to develop breast cancer? Well, we know that most cancers do not result from a single mutation. Let us assume that we know a series of mutations that can lead to breast cancer. If an individual inherits from either parent, through the egg or the sperm, one of the mutations in the series, there is one less mutation that has to occur within the breast cells of that individual for cancer to arise. The mutation inherited from the parent is present in all of the individual's cells, including the breast cells. Having inherited one of the mutations in the series increases the likelihood that breast cancer will develop, but does not make it inevitable. For this reason, many cases of inherited cancer tend to occur in younger individuals. Sporadic cancers often occur in older adults, as it takes a longer time to accumulate the mutations necessary to transform a normal cell into a cancerous one.

Much attention has been given to the discovery of "the genes for breast cancer." In fact, what was discovered was two genes, *BRCA1* and *BRCA2*, which are now classified as tumor suppressors. These are not the only two genes involved in the development of breast cancer. However, inheriting a mutated form of either of these two genes greatly increases an individual's risk of developing breast cancer. It makes the most sense to refer to *BRCA1* and *BRCA2* as breast cancer susceptibility genes. In the United States, approximately 1 in 12 women develops breast cancer. However, only between 5% and 10% of breast cancers are hereditary. Another way of saying this is that between 90% and 95% of breast cancer patients do not carry germ line mutations that predispose them to developing the disease. In these patients, all of the mutations necessary to develop breast cancer occurred sporadically within breast cells sometime during their lifetime. In the other 5% to 10% of patients with hereditary cancer, at least one of the mutations necessary for cancer development was present at birth.

The two genes were discovered by examining families with a very high incidence of early onset breast cancer. The BRCA1 gene was mapped to a position on chromosome 17 and the BRCA2 gene to a position on chromosome 13. Both of these genes have been shown to encode proteins that are needed for the repair of DNA damage. Therefore, if either or both are missing or not functioning correctly, damaged DNA may not be effectively repaired. In these individuals, DNA damage will accumulate more rapidly and accelerate cancer formation.

The characterization of these genes points out another important aspect of cancer research. In families with inherited cancer, it is now possible to do genetic testing to determine just which individuals carry mutations in genes such as BRCA1 and BRCA2 that are known to increase the likelihood of cancer. These tests allow individuals with high levels of cancer risk to be identified. They can then have checkups more frequently and improve the chances of detecting any cancer that does arrise early.

SIDE EFFECTS AND DIRTY TRICKS

Drug Resistance

As cancer cells accumulate mutations, some of the changes they cause contribute to the severity of the disease and the difficulty of its treatment. Cancer cells can ultimately become resistant to chemotherapy and spread even after aggressive treatment. This kind of drug resistance has been shown to result from mutations that occur in tumor cells. There are several kinds of mutation that can confer resistance to drugs, but two examples are especially dangerous in cancer patients. The first type of mutation causes increased expression of a protein called P-glycoprotein. This is a transmembrane protein that can function as a pump to eject cancer drugs from the cell before they can have therapeutic effects. As expression of the P-glycoprotein protein increases in cancer cells, they increase their capacity to pump chemotherapy drugs out, and this dramatically decreases treatment effectiveness.

The second drug resistance feature shared by many cancer cells centers around the anti-apoptotic character of cancer. Many chemotherapy agents use the strategy of damaging DNA and rely on the DNA damage detection system in cells to activate apoptosis. However, many cancer cells experience mutations in these very systems during their development. They may no longer be able to detect DNA damage or use signals from DNA damage detection to activate apoptosis. This represents a central paradox of cancer therapy: The very characteristics that are defined as hallmarks of cancer cells also defeat the treatment strategies commonly used in cancer therapy.

Cachexia

One of the life-threatening side effects of many kinds of cancer is uncontrollable weight loss, specifically marked by the wasting away of muscle and other lean tissue. This condition, termed *cachexia*, is the direct cause of 10–20% of cancer deaths. The causes of cachexia are complex and not fully understood. It is clear, however, that the metabolism of healthy cells is affected by signals released by cancer cells. Some of the signals released include proteins and lipids that affect appetite control centers in the brain. Other signals can directly induce metabolic changes within healthy cells. These metabolic changes then lead to loss of healthy cells and the wasting of healthy tissue. The severe wasting of cachexia is not the consequence of the diversion of part of the blood supply to tumors due to the overgrowth of blood vessels and the angiogenic properties of cancer cells. Tumor-derived signals released in the bloodstream are received by muscle and change the rate of protein degradation within muscle cells. This capacity of tumor cells to elude intrinsic damage monitoring systems, avoid cell death, hijack the blood supply, resist drug therapy, and cause the deterioration of healthy tissue indicates the severity of the disease and the imposing challenges of developing treatments.

Aneuploidy

Most cancer cells exhibit **aneuploidy**—that is, they have the incorrect number of chromosomes (either too many or too few) (Figure 11). Most normal human somatic cells have 46 chromosomes—cancer cells can sometimes have more than 100 chromosomes. This abnormal amount of chromosomes can occur when cell division is not correctly performed; it may also be triggered when DNA-damaging agents (such as radiation) cause chromosomes to break and become rearranged. Once chromosomes are in pieces, it may be difficult for the cell cycle machinery to correctly divide chromosomes into daughter cells during cell division.

Whether aneuploidy is the cause or the consequence of cancers has been widely debated. However, there is mounting evidence that having the incorrect number of chromosomes promotes a cancerous state. Specific tumor types have been shown to have specific duplications of particular chromosomes, and mutations in specific genes that help regulate the correct segregation of chromosomes into daughter cells upon cell division have been identified in cancers. The presence or absence of entire chromosomes causes the cell to have incorrect numbers of thousands of genes, which is more devastating to a cell than a few random mutations that accumulate over time.

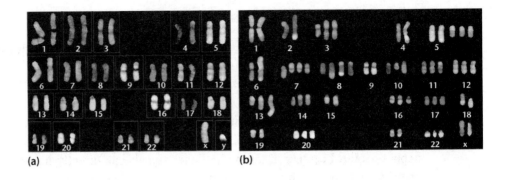

FIGURE 11: Karyotypes of (a) a Normal Cell and (b) a Cancer Cell. Note that in the cancer cell karyotype there are several examples of aneuploidy. For example, chromosomes 3, 7, 8, 13 and others. The staining technique used to obtain these karyotypes allows each different chromosome to be recognized because it stains a different color but because this is a black and white image these color differences are not apparent. Such color differences also reveal many other abnormal characteristics of the cancer cell karyotype including deletions and swapped DNA (translocations).
Source: National Institute of Health Genetics/Ried 8340061

Differentiation and Cancer Stem Cells

During development of an organism, cells become **differentiated**, meaning that they become mature cells of a particular type with a particular function. All cells arise from embryonic stem cells, which are capable of both dividing indefinitely and becoming any cell in the body. Over time, stem cells will begin to take on specific characteristics of a particular tissue. For example, a stem cell destined to become a neuron (nerve cell) will start to take on the characteristic shape of a neuron and will begin producing proteins specific to the function of neurons. Cell division typically stops as the cell matures to its final form. In an adult, there are small amounts of stem cells found throughout the body, which are needed to provide cells to replace cells that die or become damaged. **Dedifferentiation** describes a process in which a mature, specific type of cell reverts back to an earlier stage in its development. Such cells do not carry out the particular function of any tissue, but are capable of limitless division. One theory of cancer development is that significant mutations occur in cancerous cells that cause them to undergo dedifferentiation.

Alternatively, there is a theory that cancer may arise from **cancer stem cells (CSCs).** Stem cells have the capability of dividing and forming populations of cells that can mature into different cell types. CSCs could arise when normal

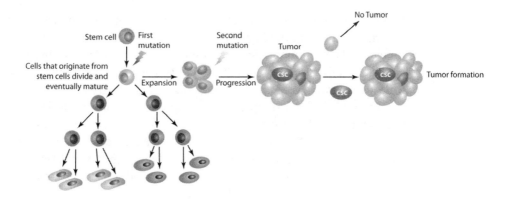

FIGURE 12: Tumors May Arise from Cancer Stem Cells (CSCs). If an immature stem cell becomes mutated and then the mutated daughter cells receive at least one more key mutation, a cancerous tumor can result. Some cells of this tumor would be composed of CSCs, capable of limitless division. Some tumor cells may break free from the original tumor, but may not survive chemotherapy or efficiently metastasize. It is believed that CSCs are efficient at forming new, additional tumors. Traditional cancer treatments may effectively destroy most cells of the tumor, but fail to destroy the CSCs, which are more resistant to chemotherapy. These cells may survive the chemotherapy treatments and allow the tumors to reoccur.

adult stem cells become mutated and experience excessive growth (Figure 12). The presence of such stem cells in a tumor could help give rise to the many diverse types of cells that are often found in solid tumors. CSCs may make up only a small percentage of a tumor. It is believed that they may be more resistant to chemotherapy, possibly due to the enhanced presence of drug-resistant proteins like those mentioned earlier. The presence of small numbers of CSCs that can thus survive chemotherapy, even when there is substantial shrinkage of tumors, may explain cancer reoccurrence. Increased research is being conducted on the ability to identify and specifically target CSCs during treatment, in the hope that metastases and reoccurrence can be prevented.

Failure of the Immune System and Cancer

Individuals with weakened immune systems (for example, AIDS patients and those with suppressed immune systems due to organ transplants) have a higher risk of developing cancer. This suggests that having a normally functioning immune system may provide some protection against cancer. The purpose of the immune system is to protect the body from foreign invaders (including bacteria and viruses) and abnormal cells (including cancer cells), which need to be safely eliminated. The immune system accomplishes this protection through two major classes of cells: B cells and T cells. **B cells** are antibody-producing cells. **Antibodies** are proteins that recognize and bind to molecules that are unrecognizable—for example, foreign proteins from bacteria or viruses, or proteins that may be incorrectly made on the surface of cancer cells. Once tagged with an antibody, the bacterium, virus, or cancer cell has been identified for destruction by other immune system cells that eliminate antibody-tagged cells. Alternatively, some cancer cells may be destroyed by **T cells**—more specifically, cells that are called **cytotoxic T lymphocytes**. These cells recognize abnormal proteins on the surface of cancer cells and directly destroy them. When cancer cells start to acquire the many mutations that allow their transformation, it is likely that they also start to accumulate incorrectly made proteins on their surface.

The B cells and T cells of the immune system, which circulate throughout the body and are on constant "surveillance" for invaders, should be able to immediately identify these mistaken proteins and destroy the problematic cells that are producing them. When a cancerous tumor occurs, this indicates that the immune system has somehow "missed" destroying these abnormal cells. This may happen for one of several reasons. It is known that some cancer cells produce proteins that may inhibit the activation or proliferation of T cells needed to fight the cancer. In other cases, cancer cells may not produce faulty proteins on their surface, where they can be readily detected by immune system cells. The cells of the immune system need to be able to trigger apoptosis in the cancerous cells in order to eliminate them. As previously discussed, cancer cells can accumulate mutations early in their transformation that cause them to be resistant to apoptosis. This may also allow them to escape destruction by immune system cells.

Interestingly, there is also evidence that chronic inflammation, caused by a constantly overactive immune system, can increase predisposition to cancer. Chronic inflammation may be caused by continuous infection with bacteria or viruses or by autoimmune diseases (diseases in which antibodies are incorrectly made to work against normal body tissues, like what occurs in rheumatoid arthritis). Cells of the immune system often produce free radicals in order to damage the DNA of invading/targeted cells, triggering apoptosis. These free radicals may also, unfortunately, damage the DNA of healthy cells in the vicinity of the infection, resulting in mutations that may lead to cancer formation. Cells of the immune system release proteins that trigger angiogenesis to help rebuild damaged tissues. However, such activity can also, unfortunately, help promote the growth of tumors. Tissues that have prolonged exposure to immune responses therefore sustain collateral damage that can lead to cancer. For example, patients with inflammatory bowel disease have been shown to have increased risk for colon cancer.

Epigenetics and Cancer

The term **epigenetics** refers to inherited changes in gene expression, not including changes in the DNA sequence itself. Even though every cell in the body contains the same genes (estimated at approximately 20,000), not every gene is used to make protein in every cell at all times. There are mechanisms that control which genes are able to be used to make proteins. These mechanisms include DNA methylation and histone modification, and they are considered epigenetic, since both modifications occur to the DNA, but they do not alter the DNA sequence itself, as a mutation does. These changes are also able to be inherited: If a particular gene is "turned off" by epigenetics in parents, it is likely that the same gene will be controlled to be turned off in offspring.

DNA methylation is the addition of a group of atoms (a methyl group, containing carbon and hydrogen) to the DNA sequence. Methylated regions of DNA cannot be used to make proteins, as the methyl group makes the DNA inaccessible to the enzymes necessary to allow genes to be used to make proteins. Similarly, **histones** are proteins that help keep DNA folded, packaged, and organized inside the nucleus of a cell. When the histones are arranged to keep the DNA sequence tightly compacted (Figure 13), the DNA cannot be accessed to allow expression of the genes to form protein. The tightness of the arrangement of the histones is controlled by the presence or absence of chemical tags, such as acetyl or methyl groups. For example, the addition of acetyl groups to histones tends to loosen their association with DNA, allowing the DNA to be used for expressing genes into proteins.

It is believed that, in addition to mutations in the DNA sequence, faulty epigenetic changes to the DNA may contribute to cancer. For example, turning a tumor suppressor gene off through the modification of histones can incorrectly allow uncontrolled growth. This would have the same devastating effect as a mutation of the tumor-suppressing gene that prevents its activity. It is believed that aging, diet, and environmental factors may all influence the degree of methylation of DNA. Thus, it is possible that epigenetic changes occur over the lifespan of an individual and can contribute to cancer formation or promotion. To better study the role of epigenetics in cancer, the National Institutes of Health (NIH) has started The Cancer Genome Atlas, in order to compare both mutations and epigenetic changes in cancer cells, and is available to the public (www.cancergenome.gov).

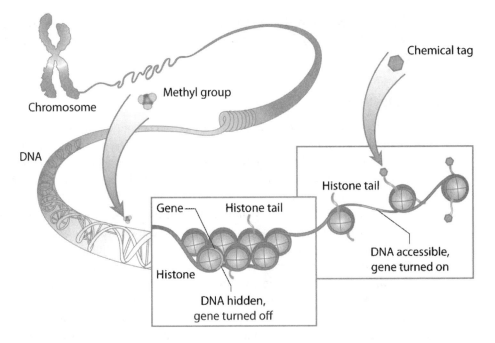

FIGURE 13: Epigenetics and Gene Expression. DNA that is methylated or that is highly compacted due to the tight arrangement of histone proteins cannot be accessed and used for gene expression. The tightness of the association of DNA with histones can be controlled by the addition of chemical tags to the histones. These epigenetic modifications to DNA can be inherited.
Source: http://www.genome.gov/27532724.

A new movement in the area of epigenetics is the study of **microRNA (miRNA)**. When genes are being expressed, the first step is to use the DNA sequence to make a messenger RNA (mRNA), which is then used as the final blueprint to make a protein. miRNAs have been shown to block mRNA, so that it cannot be used to make proteins. Some cancer cells have been shown to have aberrant levels of miRNAs, which in turn, can incorrectly turn "off" expression of needed genes. The identification of specific miRNAs that may be incorrectly present in cancerous cells will be a new avenue for cancer treatments.

CANCER DIAGNOSIS

As mentioned previously, "cancer" is a term used to describe over 100 different diseases characterized by uncontrolled growth of cells. Because of the diversity of cancer, there is diversity in how the disease is recognized in patients. However, there are some general cancer "warning signs" and some standard diagnostic tests that are frequently utilized. The most common diagnostic tests will be mentioned in this segment. For some cancers, the symptoms seem so "vague" that early warning signs are often missed by patients. As seen in our example, Rita had some mild indigestion, but did not recognize this as a potential cancer symptom.

Warning signs:

- unexplained weight loss
- persistent or intermittent fever
- fatigue
- sores that do not heal
- changes in bowel or bladder habits
- pain
- indigestion or trouble swallowing
- thickening or lumps, or changes in any moles or warts
- skin changes: darkening, yellowing, redness or itching
- nagging cough or hoarseness
- unusual bleeding or discharge

Common Diagnostic Tests

Blood Tests

Blood tests are often performed to look for **tumor markers**—proteins that are found in elevated levels in certain cancers. However, elevated levels of these proteins may not *always* indicate cancer (sometimes other "normal" or less serious conditions alter their levels), so additional testing is required. The levels of tumor markers can also be followed during treatment to gauge how well the patient is responding to therapy. Common proteins elevated in cancer are listed in Table 3.

TABLE 3: Some Common Tumor Markers and The Cancers That They Are Used to Diagnose.

Tumor Marker	Cancer	Description
CA-125 (Cancer antigen-125)	Ovarian	Protein produced when certain abdominal tissues are inflamed
PSA (Prostate specific antigen)	Prostate	Protein produced by prostate cells—when too many cells are present, levels of this protein may increase
AFP (alpha-fetoprotein)	Liver	Protein produced by liver tumors, but also elevated in hepatitis (liver infection)
Calcitonin	Medullary thyroid carcinoma (MTC)	This is a hormone produced by thyroid, and is increased in MTC, a rare thyroid cancer
Bcr-abl	Chronic myeloid leukemia (CML)	Abnormal protein only produced in CML

Imaging Tests

Imaging tests use X-rays, ultrasound, or magnetic fields to create images of internal structures, so that any potential tumors can be visualized. **CT (computed tomography) scans** use multiple X-ray images that are synthesized by a computer to form three-dimensional images. In addition to the initial identification of the presence of a tumor, CT scans are often repeated throughout treatment to follow the size and presence of tumors in order to evaluate the effectiveness of treatment. **Magnetic resonance imaging (MRI)** uses magnetic fields, not radiation, to create images of internal structures. It is often used when cancer is suspected in soft tissues (for example, brain, spinal cord, or muscle) and larger portions of the body need to be examined.

Biopsy

For many cancers, after a tumor is detected, a biopsy is performed to remove a small piece of the tumor from the patient. Experts in recognizing cancer cells look at the tumor cells from the biopsy using a microscope. The tumor cells may also be examined for the presence of specific tumor markers or DNA mutations. The overall size and shape of cancer cells are abnormal, and cancer cells often tend to be arranged together in an abnormal fashion Figure 14. The identification of cancers by "sight" only is not easy, and specialists will usually obtain multiple opinions on their findings, and will correlate visual results with laboratory tests of tumor markers to make a diagnosis. Increasingly, biopsy samples are used for direct testing for DNA mutations to help with cancer diagnosis.

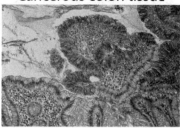

FIGURE 14: A View of "Normal" Colon Cells and Cancerous Colon Cells. The cancerous cells are more numerous, and arranged in a more disorganized fashion. *Source:* (left) Gary D. Gaugler/Photo Researchers, Inc.; (right) CNRI/Photo Researchers, Inc.

Cancer Staging

Once a cancer diagnosis is made, "staging" is used to further characterize the extent of the disease and to best evaluate treatment options. Typically, the cancer is staged by giving numbers for three different disease characteristics:

- T: A number (1–4) is given to identify the size of the tumor and if it appears to be contained in one tissue. Larger numbers indicate larger tumors and local spreading of the tumor.

- N: 0–2 is given to identify the presence of tumor cells in lymph nodes, immune system organs that tend to collect spreading cells. A larger number indicates more spreading to these regions.

- M: "1" is given to indicate metastases (spreading of the tumor to other locations); a tumor that has no evidence of spreading is considered "0".

Based upon these separate characteristics, an overall number (I, II, III, or IV) is assigned to the cancer, with IV being the most severe staging level. For example, a "T1N0M0" tumor would be stage I, while a "T4N2M1" tumor would be classified as stage IV. Staging is critical in that tumors that are stage I are usually contained in one location, and may be more treatable using surgery than cancers that have reached stage IV. In such cases, more widespread treatments, such as chemotherapy, may be chosen as a first step.

MOST COMMON CANCER TREATMENTS

Once a patient has been diagnosed with cancer, the first questions are usually related to the prognosis ("how bad is it?") and how it can be treated. Treatments have become highly individualized, and the suggested course of treatment depends on many factors, including the general health of the patient (how would he or she tolerate specific treatment paths), the location of the tumor, and the extent that the tumor has metastasized. Overall, the goals of treatment are to remove as much of the cancer as possible, and then to kill or severely inhibit the growth of remaining tumor cells. The three most common cancer treatments include surgery, radiation, and chemotherapy.

Surgery

Surgery is performed to remove (completely or as much as possible) solid cancerous tumors. With the exception of blood cancers such as leukemia, many cancers are due to tumors that form in the body. Depending on the cancer type or location, however, surgery is sometimes not the first course of treatment. In certain cancers, such as a specific type of breast cancer called inflammatory breast cancer (IBC), it has been found that "shrinking" the tumor first with other types of treatments prior to surgery leads to better outcomes. In addition, sometimes the location of tumors makes them inaccessible, or dangerous to remove surgically, in which case alternative treatment methods will be tried first. As with pulling weeds from a garden removing cancer surgically can be difficult, and sometimes the "roots" of the cancer can be left behind. Sometimes it is clear after surgery that the entire cancer has been removed, however, if there is any doubt that cancer cells remain, additional treatments, such as chemotherapy or radiation, are often suggested.

Radiation

Radiation therapy is the use of high energy waves (radiation) to kill, or inhibit the cell division, of cancer cells. Radiation treatments can be given in the two forms: either an external beam of radiation can be aimed at the tumor, or alternatively, a "seed" of radioactivity can be surgically implanted within the tumor. The amount of radiation used and the duration of treatments vary, depending upon the type of tumor, location, and the health of the patient. As was mentioned previously, radiation causes DNA damage, and is also one factor that can also lead to cancer itself. As a cancer treatment, radiation is used to cause DNA damage in a tumor. Since the tumor cells are rapidly dividing, the DNA damage caused by radiation should prevent them from being able to replicate their DNA correctly during cell division, triggering apoptosis to occur in these cells. Unfortunately, "normal" cells may be damaged by radiation treatments as well, so the tumor must be carefully targeted to prevent such collateral damage as much as possible. Additionally, some cancers are known to respond better to radiation treatments than others, and large solid tumors are also difficult to treat with this method. In some instances, where surgery is dangerous or the location of the tumor makes surgery impossible, radiation is a good alternative first treatment. In addition, radiation is also sometimes used to relieve pain by shrinking tumors when metastasis is extensive and the ability to perform surgery to relieve pain is limited.

Chemotherapy

Chemotherapy is the term given to broadly describe the use of drugs to treat cancer. A wide range of drugs is now available, and as individual types of cancers continue to be better understood, more drug cocktails tailored to individual patients will continue to be developed. These new therapies tend to specifically target individual mutated pathways in cancer cells, allowing for fewer side effects. In general, traditional chemotherapy drugs tend to target cells that are rapidly dividing. Unfortunately, there are some healthy cells (like the cells that line the digestive track and hair follicles) that also divide rapidly, and can be harmed by chemotherapy. This can lead to some side effects (nausea, hair loss). There are many drugs today that can be given to counteract some of these side effects, which makes modern chemotherapy easier to tolerate than the last generation of cancer treatments. Such treatments usually target one of three mechanisms of rapid cell division: (1) they damage DNA, which disrupts DNA replication, (2) they interfere with the process of DNA replication itself, or (3) they directly interfere with the process of cell division. Some of these drugs are given orally, and can be taken at home, while others require intravenous infusion, and can be given while in the hospital or at an outpatient treatment facility. Some representative chemotherapy drugs are described in Table 4.

OTHER CANCER TREATMENTS

Bone Marrow Transplantation

Leukemias and lymphomas are two of the most common types of cancer that can be treated with bone marrow transplantation. Bone marrow contains the stem cells that are capable of producing blood cells, and marrow is often destroyed due to the chemotherapy and radiation used to treat these types of cancers. Typically,

TABLE 4: Some Representative Cancer Chemotherapy Drugs.

Drug Name	Mechanism	Some Representative Cancers Treated
Cisplatin (Platinol)	Causes DNA damage	Ovarian, head and neck, esophageal, lung
Doxorubicin (Adriamycin)	Causes DNA damage	Lung, breast, leukemia, bladder, thyroid
5-Fluorouracil (Efudex, Adrucil)	Inhibits DNA replication	Skin cancer
Methotrexate (Trexall)	Inhibits DNA replication	Leukemia, lymphoma, bladder
Paclitaxel (Taxol)	Inhibits mitosis	Ovarian, breast, lung
Vinblastine (Velban)	Inhibits mitosis	Breast, testicular, lymphoma
Vincristine (Oncovin)	Inhibits mitosis	Hodgkin's disease, lymphoma, neuroblastoma

marrow can be taken from a patient after extensive chemotherapy and radiation, which should destroy most (or, hopefully, all) cancer cells. This marrow can then be re-introduced when needed. Alternatively, donor marrow from a healthy individual (preferably a family member that closely "matches" the genetic makeup of the patient) can be given. Successful transplants allow new blood cells to restore a functioning immune system in the patient, which had been destroyed by treatment.

Targeted Therapies

In recent years, great advances have been made in understanding the individual mutations that occur in specific cancers. This has led to a new surge in cancer drug development: therapies targeted to individual mutations found in specific types of cancer. Scientists also refer to this approach as **pharmacogenomics**—customized medicine based on a person's genetics. One benefit of this type of treatment is that a molecular pathway is targeted that is only defective in cancer cells—this allows for fewer side effects as "normal" cells should not be affected by such treatments. However, since so many different mutations act in combination to make a cell cancerous, blocking the effect of one individual mutation is often not enough to completely destroy the cancer. For this reason, many of these targeted therapy drugs are given in combination with traditional chemotherapy. As more mutations involved in individual cancers are identified through the Cancer Genome Atlas Project, it is hoped that the major pathways that are disrupted by mutations in a variety of cancers will be identified.

TABLE 5: Some Representative Drugs Used for Targeted Therapy.

Drug Name	Mechanism	Representative Cancer Treated
Erlotinib (Tarceva)	Blocks the activity of a growth factor receptor that is over-active	Lung, pancreatic
Bevacizumab (Avastin)	Blocks VEGF—serves as an angiogenesis inhibitor	Colorectal
Tamoxifen (Nolvadex)	Blocks the activity of estrogen receptors, which are overactive in some cancers	Breast
Ipilimumab (Yervoy)	Increases activity of immune cells	Melanoma
Imatinib (Gleevec)	Blocks the activity of an enzyme needed for cell division	Leukemia, gastrointestinal

Nevertheless, the currently approved targeted therapies have greatly increased the survival rates for certain cancers. Imatinib, marketed as "Gleevec," was one of the first targeted therapies. This drug is an inhibitor of an enzyme that is overactive in a specific cancer called chronic myelogenous leukemia (CML). Prior to the introduction of this drug, very few patients survived for five years. Currently, there is roughly an 89% survival rate for CML due to the use of this new treatment. Gleevec has now also been approved for other cancers that contain the same enzyme mutation. Some additional examples of targeted therapies are given in Table 5.

Tumor Vaccines

Vaccination, in general, works by stimulating the immune system to "fight" a specific invader. Vaccines have traditionally been given to prevent infectious diseases, and one anti-cancer vaccine, Gardasil (described earlier), has been approved for the prevention of cervical cancer in individuals who have not been previously infected with HPV. However, there is now also widespread use of vaccines as a cancer treatment—the goal of such vaccination is to increase the immune system's response against cancer. Tumor cells that have been killed, or parts of tumor cells, can be used as a vaccine and directly injected back into the patient. Such direct challenge of the patient's immune system should greatly increase the response of the immune system to the actual tumor cells that are present. Currently, vaccine treatments are in clinical trials for both melanoma and prostate cancer.

Gene Therapy

Gene therapy is the introduction of new DNA into cells in order to allow proteins to be made that are needed to overcome a disease. This may involve adding missing or mutated genes, such as the gene for p53, back into tumor cells. Alternatively, gene therapy may be used to introduce genes that can help trigger apoptosis, or genes that may block angiogenesis. Currently, there are over 1400 gene therapy clinical trials that target cancer.

Clinical Trials

Clinical trials are used to study the safety and effectiveness of new cancer treatments, as well as new methods of prevention and detection of cancer. In the United States, the FDA oversees the clinical trial regulatory process for new medicines and medical devices. FDA clinical trials are designated as *Phase I*, *Phase II*, or *Phase III* depending upon their goals. Phase I trials are designed to simply study the safety of a new drug, and to understand how the dosing of the drug can be given. Phase II trials begin to examine how effective the drug is against a specific type of cancer, while Phase III trials are large-scale trials that compare the new drug with the currently accepted treatments. For cancer trials, new drugs are often given in combination with existing treatments—experimental drugs are not used as the only treatment during the trial.

Many cancer patients may turn to clinical trials when their current treatments do not seem to be effectively working. Fifty-nine hospitals in the United States. are also designated as "Cancer Centers" by the National Cancer Institute (NCI), which means that they have received federal grant funding to perform research in cancer prevention, diagnosis, and treatments. A list of the designated hospitals can be found on the NCI website. Patients treated at Cancer Centers, or at medical university teaching hospitals that conduct research, may also be recruited for on-going clinical trials by their treating physicians at the time of their diagnosis. However, not all patients are eligible to participate in all clinical trials. Each specific trial will have guidelines for the age, type of cancer, overall health, and previous treatments given that will determine if a patient can be enrolled in a trial. Likewise, if a patient's cancer progresses while enrolled in a trial, or if serious side effects of the trial drug are encountered, they are usually removed from participation. Information about specific clinical trials can be found at www.clinicaltrials.gov and at the National Cancer Institute website (www.cancer.gov/clinicaltrials). These websites allow specific searches of cancer types, and give information on trial designs and locations. Progress in developing new, effective cancer treatments is slow, however, some patients have been able to achieve "stable disease"—they are chronically living with cancer for many years due to recent advances in treatment.

CONCLUSION

Cancer will undoubtedly remain one of the leading causes of death in the United States for years to come. However, each decade brings more advances in cancer prevention, early detection, and treatment. More and more patients, like Rita discussed in the introduction, are surviving their initial cancer diagnosis due to earlier detection and better treatments, and living comfortably for many years. Patients are also becoming better informed, and more educated about their conditions and are therefore more able to be proactive about their own treatment options. As more targeted therapies are developed with fewer overall side effects, cancer chemotherapy will continue to improve and become more easily tolerated by patients. Since there is now a better understanding of the many types of mutations that can contribute to cancer, the focus has switched from finding an overall cancer cure to looking at each tumor individually and specifically targeting the unique defects found in that one individual tumor.

I hope that this booklet has provided you with a greater understanding of the cancer causes, diagnosis methods, and treatments that may have touched the lives of your families and friends. As the field of cancer research is continuously progressing, the web resources provided at the end of this booklet will serve as an excellent source for updates. Please feel free to contact me at the address below if you have any comments or thoughts about this booklet or advances in cancer cell biology.

Dorothy Lobo, Ph.D.
Associate Professor
Monmouth University
Biology Department
400 Cedar Avenue
West Long Branch, NJ 07764
E-mail: dlobo@monmouth.edu

WEB RESOURCES

American Cancer Society (www.cancer.org): Information is provided on the biology of cancer and cancer prevention, treatment, and research.

The Cancer Genome Atlas (http://cgap.nci.nih.gov/): This initiative of the National Cancer Institute and the National Human Genome Research Institute focuses on the use of genome analysis techniques to improve our understanding of the molecular basis of cancer. It provides a resource for learning about the genetics of different types of cancer.

Cancer.Gov (www.cancer.gov): Created by the National Cancer Institute of the National Institutes of Health, this website contains straightforward information about cancer intended for patients, health-care providers, and the public.

CancerQuest (www.cancerquest.org): This education and outreach website, maintained by Emory University, provides comprehensive, yet easy to read, information about the biology of cancer and cancer treatments.

ClinicalTrials.gov (www.clinicaltrials.gov): Maintained by the National Institutes of Health, this website contains a searchable database of all federally and privately supported clinical trials worldwide.

Inside Cancer (www.insidecancer.org): This resource for students and educators is divided into four sections: hallmarks of cancer, cancer causes and prevention, cancer diagnosis and treatment, and pathways to cancer. Animations, commentaries, and lesson planning ideas for teachers to align content to national education standards.

MD Anderson Cancer Center (www.mdanderson.org): The MD Anderson Cancer Center, at the University of Texas, is one of the leading cancer treatment centers in the United States. Its website contains information on cancer prevention, treatment, and clinical trials.

National Cancer Institute Clinical Trials (www.cancer.gov/clinicaltrials): This website provides a search engine for the thousands of clinical trails run by the National Cancer Institute and currently accepting patients.

OncoLink (www.oncolink.org): Maintained by the University of Pennsylvania Cancer Center, this website contains basic information on cancer types, treatments, and support.

BOOKS AND ARTICLES

Brenner, D. J., & Hall, E. J. (2007). Computed tomography—an increasing source of radiation exposure. *New England Journal of Medicine, 357,* 2277–2284.

Cancer prevention. (2011). *Nature Outlook* [Supplement]. *Nature, 471,* S1–S24.

Esteller, M. (2011). Epigenetic changes in cancer. *Scientist, 25,* 34–39.

Gibbs, W. W. (2003). Untangling the roots of cancer. *Scientific American, 289,* 30–39.

Hanahan, D., & Weinberg, R. A. (2000). The hallmarks of cancer. *Cell, 100,* 57–70.

Hanahan, D., & Weinberg, R. A. (2011). Hallmarks of cancer: The next generation. *Cell, 144,* 646–674.

Heath, J. R., Davis, M. E., & Hood, L. (2009). Nanomedicine targets cancer. *Scientific American, 300*(2), 44–51.

Kaiser, J. (2008). A detailed genetic portrait of the deadliest human cancers. *Science, 321,* 1280–1281.

Ledford, H. (2011). The cancer genome challenge. *Nature, 464,* 972–974.

Manchado, E., & Malumbres, M. (2011). Targeting aneuploidy for cancer therapy. *Cell, 143,* 465–466.

Pecorino, L. (2008). *Molecular biology of cancer: Mechanisms, targets, and therapeutics.* Oxford University Press, Oxford, UK.

Rothenberg, M. E., Clarke, M. F., & Diehn, M. (2010). The myc connection: ES cells and cancer. *Cell, 143,* 184–186.

Stratton, M. R., Campbell, P. J., & Futreal, P. A. (2009). The cancer genome. *Nature, 458,* 719–724.

Ventura, A., & Jacks, T. (2009). MicroRNAs and cancer: Short RNAs go a long way. *Cell, 136,* 586–591.

Von Hofe, E. (2011). A new ally against cancer. *Scientific American, 305*(4), 66–71.

Visvader, J. E. (2011). Cells of origin in cancer. *Nature, 469,* 314–322.

Weinberg, R. A. (1996). How cancer arises. *Scientific American, 275*(3), 62–70.

Weinberg, R. A. (2006). *The biology of cancer.* New York: Garland Science.